HOW TO PASS

# FLASH REVISE!
## HIGHER
# MATHS

Brian Logan

D1340229

HODDER
GIBSON
AN HACHETTE UK COMPANY

Orders: please contact Bookpoint Ltd, 130 Milton Park, Abingdon, Oxon OX14 4SB. Telephone: (44) 01235 827720. Fax: (44) 01235 400454. Lines are open 9.00–5.00, Monday to Saturday, with a 24-hour message answering service. Visit our website at www.hoddereducation.co.uk. Hodder Gibson can be contacted direct on: Tel: 0141 848 1609; Fax: 0141 889 6315; email: hoddergibson@hodder.co.uk

© Brian Logan 2011

First published in 2011 by
Hodder Gibson, an imprint of Hodder Education,
An Hachette UK Company
2a Christie Street
Paisley PA1 1NB

Impression number    5  4  3  2
Year    2013  2012

Cover photo © Nikreates/Alamy; © iStockphoto.com
Typeset in Gill Sans Alternative by GreenGate Publishing, Tonbridge, Kent
Printed in Great Britain by CPI Group (UK) Ltd, Croydon, CR0 4YY

A catalogue record for this title is available from the British Library

ISBN: 978 1444 120479

# Contents

# Revision (1)

**Q1** Expand the brackets and simplify $(3x + 2y)^2$.

**Q2** Expand the brackets and simplify $(2x - 5)(x + 3)$.

**Q3** Expand the brackets and simplify $(x + 2)(x^2 - 3x + 2)$.

**Q4** Factorise $8x^2 - 6x$.

**ANSWERS** ⟩⟩

**1** $(3x + 2y)^2 = (3x + 2y)(3x + 2y)$

$\qquad\qquad = 9x^2 + 6xy + 6xy + 4y^2$

$\qquad\qquad = 9x^2 + 12xy + 4y^2$

**2** $(2x - 5)(x + 3) = 2x^2 + 6x - 5x - 15$

$\qquad\qquad\quad = 2x^2 + x - 15$

**3** $(x + 2)(x^2 - 3x + 2) = x^3 - 3x^2 + 2x + 2x^2 - 6x + 4$

$\qquad\qquad\qquad\quad = x^3 - x^2 - 4x + 4$

**4** $8x^2 - 6x = 2x(4x - 3)$

# Revision (2)

**Q1** Factorise $9x^2 - 25y^2$.

**Q2** Factorise $3a^2 - 5a + 2$.

**Q3** Factorise $2\cos^2 x - 3\cos x - 2$.

**Q4** Solve the system of equations:
$$3x - y = 13$$
$$5x + 2y = 18.$$

**ANSWERS** ▶▶

1  $9x^2 - 25y^2 = (3x + 5y)(3x - 5y)$

2  $3a^2 - 5a + 2 = (3a - 2)(a - 1)$

3  $2\cos^2 x - 3\cos x - 2 = (2\cos x + 1)(\cos x - 2)$

4  Double the equation $3x - y = 13$, then add to the equation $5x + 2y = 18$. This leads to $11x = 44$ and hence $x = 4$, $y = -1$.

# Revision (3)

**Q1** Write down the equation of the straight line passing through the point $(0, c)$ with gradient $m$.

**Q2** Simplify $\dfrac{3}{x} - \dfrac{5}{x-1}$.

**Q3** Simplify $\dfrac{3x + 15}{x^2 + 6x + 5}$.

**Q4** Simplify $\dfrac{3a}{5} \div \dfrac{9}{10b}$.

**ANSWERS** ▶▶

**1** $y = mx + c$

**2** $\dfrac{3}{x} - \dfrac{5}{x - 1} = \dfrac{3(x - 1) - 5x}{x(x - 1)}$

$= \dfrac{3x - 3 - 5x}{x(x - 1)} = \dfrac{-2x - 3}{x(x - 1)}$

**3** $\dfrac{3x + 15}{x^2 + 6x + 5} = \dfrac{3(x + 5)}{(x + 1)(x + 5)}$

$= \dfrac{3}{(x + 1)}$

**4** $\dfrac{3a}{5} \div \dfrac{9}{10b} = \dfrac{3a}{5} \times \dfrac{10b}{9}$

$= \dfrac{30ab}{45} = \dfrac{2ab}{3}$

## Revision (4)

**Q1** Simplify $\sqrt{72}$.

**Q2** Express $\dfrac{8}{\sqrt{2}}$ with a rational denominator.

**Q3** Expand the brackets and simplify $p^{\frac{1}{2}}(p^{\frac{1}{2}} + p^{-\frac{1}{2}})$.

**Q4** Evaluate $8^{\frac{2}{3}}$.

**ANSWERS ▶▶**

**1** $\sqrt{72} = \sqrt{36 \times 2} = 6\sqrt{2}$

**2** $\dfrac{8}{\sqrt{2}} = \dfrac{8}{\sqrt{2}} \times \dfrac{\sqrt{2}}{\sqrt{2}}$

$= \dfrac{8\sqrt{2}}{2} = 4\sqrt{2}$

**3** $p^{\frac{1}{2}}(p^{\frac{1}{2}} + p^{-\frac{1}{2}}) = p^{\frac{1}{2}} \times p^{\frac{1}{2}} + p^{\frac{1}{2}} \times p^{-\frac{1}{2}}$

$= p^{\frac{1}{2}} + p^0 = p^3 + 1$

**4** $8^{\frac{2}{3}} = (\sqrt[3]{8})^2 = 2^2 = 4$

***Exam* tip:** It is most important that you are comfortable doing the revision topics mentioned in this section. If you have difficulty with any of them, ask for advice.

# The straight line (1)

**Q1** What is the gradient $m$ of the line joining the points $(x_1, y_1)$ and $(x_2, y_2)$?

**Q2** Find the equation of the line passing through the point $(0, -3)$ with gradient 4.

**Q3** Find the equation of the line joining the points $(0, 2)$ and $(-4, 6)$.

**Q4** What is the equation of the line passing through the point $(a, b)$ with gradient $m$?

**ANSWERS** ⟩⟩

**1** $m = \dfrac{y_2 - y_1}{x_2 - x_1}$

**2** Using the formula $y = mx + c$, the equation is $y = 4x - 3$.

**3** $m = \dfrac{y_2 - y_1}{x_2 - x_1} = \dfrac{6 - 2}{-4 - 0}$

$\qquad = \dfrac{4}{-4} = -1, c = 2,$

so the equation is $y = -x + 2$.

**4** $y - b = m(x - a)$

# The straight line (2)

**Q1** Find the gradient and y-intercept of the line $4y - 3x = 8$.

**Q2** Find the gradient of the line joining the points $(3, -2)$ and $(5, 4)$.

**Q3** Find the equation of the line in Q2.

**Q4** Express the equation of the line with gradient 2 passing through the point $(5, 3)$ in the form $Ax + By + C = 0$.

**ANSWERS ▶▶**

**1** Rearrange into the form $y = mx + c$ to give $y = \frac{3}{4}x + 2$.

Hence the gradient is $\frac{3}{4}$ and the $y$-intercept is 2.

**2** $m = \dfrac{y_2 - y_1}{x_2 - x_1}$

$= \dfrac{4 - (-2)}{5 - 3}$

$= \dfrac{6}{2} = 3$

**3** Using $y - b = m(x - a)$ leads to $y - (-2) = 3(x - 3)$, which simplifies to $y = 3x - 11$.

**4** The equation is $y - 3 = 2(x - 5)$.

This can be rearranged to give $2x - y - 7 = 0$.

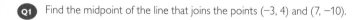

# The straight line (3)

**Q1** Find the midpoint of the line that joins the points $(-3, 4)$ and $(7, -10)$.

**Q2** Find the equation of the line that joins $(3, 1)$ and $(3, 5)$.

**Q3** Find the equation of the line that passes through the point $(-3, 8)$ and is parallel to the line with equation $2y - 6x = 9$.

**Q4** Explain why the lines $y = 3x - 5$ and $y = -\dfrac{1}{3}x$ are perpendicular.

**ANSWERS** 〉〉

**1**  The midpoint is $\left( \dfrac{-3 + 7}{2}, \dfrac{4 + (-10)}{2} \right) = (2, -3)$.

**2**  The gradient of this vertical line is undefined, so its equation is $x = 3$.

**3**  The line $2y - 6x = 9$ rearranges into $y = 3x + \dfrac{9}{2}$ and has gradient 3.

Parallel lines have the same gradient, so the required equation is

$y - 8 = 3(x + 3)$.

**4**  The product of the gradients $= 3 \times -\dfrac{1}{3} = -1$, so the lines are perpendicular.

# The straight line (4)

**Q1** A line makes an angle of 45° with the positive direction of the x-axis. What is its gradient?

**Q2** A line joins A (5, 2) and B (7, −9). Find the gradient of a line perpendicular to AB.

**Q3** Find the point of intersection of the lines with equations $y = x − 4$ and $y = 3x + 10$.

**Q4** In a triangle ABC, how would you draw the *median* from A to BC?

**ANSWERS** ⟫

1 Use the formula $m = \tan\theta$. Therefore gradient $= \tan 45° = 1$.

2 $m_{AB} = \dfrac{y_2 - y_1}{x_2 - x_1} = \dfrac{-9 - 2}{7 - 5} = -\dfrac{11}{2}$

The perpendicular gradient is $\dfrac{2}{11}$.

3 The lines intersect where $x - 4 = 3x + 10 \Rightarrow x = -7$ and $y = -11$, that is at $(-7, -11)$.

4 The median is the line from A to the *midpoint* of BC.

# The straight line (5)

**Q1** In a triangle ABC, how would you draw the *altitude* from A to BC?

**Q2** Calculate the distance $d$ between the points $(4, -1)$ and $(7, 3)$.

**Q3** Explain what is meant by the *perpendicular bisector* of a line.

**Q4** Three points A, B and C are collinear. What does this mean?

**ANSWERS ▶▶**

**1** The altitude is the line from A *perpendicular* to BC.

**2** $d = \sqrt{(x_2 - x_1)^2 + (y_2 - y_1)^2}$

$\phantom{d} = \sqrt{(7 - 4)^2 + (3 + 1)^2}$

$\phantom{d} = \sqrt{3^2 + 4^2} = \sqrt{25} = 5$

**3** The perpendicular bisector of a line is a line that passes through the midpoint and at right angles to the given line.

**4** The points lie on the same straight line ($m_{AB} = m_{BC} = m_{AC}$).

# The straight line (6)

**Q1** The line $3x - 2y - 12 = 0$ cuts the $y$-axis at C.
Find the coordinates of C.

**Q2** The points A $(3, -2)$, B $(-1, 4)$ and C $(7, q)$ are collinear. Find $q$.

**Q3** The lines with equations $by = 2x + 5$ and $y = 3x + 2$ are perpendicular. Find $b$.

**Q4** The line that joins the points S $(3, 5)$ and T $(5, k)$ has gradient 4.
Find $k$.

**ANSWERS** ⟫

**1** The line cuts the $y$-axis when $x = 0$, that is $-2y - 12 = 0 \Rightarrow y = -6$.

So C is $(0, -6)$.

**2** $m_{AB} = \dfrac{4 + 2}{-1 - 3} = \dfrac{6}{-4} = -\dfrac{3}{2}$.

$m_{BC} = \dfrac{q - 4}{7 + 1} = \dfrac{q - 4}{8}$.

Hence $\dfrac{q - 4}{8} = -\dfrac{3}{2}$.

Hence $2(q - 4) = -24$

$\Rightarrow \qquad 2q - 8 = -24$

$\Rightarrow \qquad\qquad 2q = -16$

$\Rightarrow \qquad\qquad\quad q = -8$.

**3** The gradients of the lines are $\dfrac{2}{b}$ and 3.

Hence $\dfrac{2}{b} \times 3 = -1 \Rightarrow b = -6$.

**4** $m = \dfrac{k - 5}{5 - 3} = 4$

$\Rightarrow \qquad \dfrac{k - 5}{2} = 4$

$\Rightarrow \qquad k - 5 = 8$

$\Rightarrow \qquad\qquad k = 13$

# The straight line (7)

**Q1** What does it mean when three lines are *concurrent*?

**Q2** P is the point $(-5, 2)$ and Q is the point $(3, 8)$. Find the equation of the perpendicular bisector of the line PQ.

**Q3** The gradient of a straight line is $-\sqrt{3}$. What angle does the line make with the positive direction of the $x$-axis?

**Q4** U is the point $(u, u^2)$ and V is the point $(v, v^2)$. Find the gradient of UV in its simplest form.

**ANSWERS** ▶▶

**1** The three lines pass through the same point.

**2** The midpoint of PQ is $(-1, 5)$.

$$m_{PQ} = \frac{8 - 2}{3 + 5} = \frac{6}{8} = \frac{3}{4}$$

Hence the gradient of the perpendicular bisector is $-\frac{4}{3}$

and the equation is $y - 5 = -\frac{4}{3}(x + 1)$.

**3** $m = -\sqrt{3} \Rightarrow$ angle $= \tan^{-1}(-\sqrt{3}) = 120°$

**4** $m = \dfrac{v^2 - u^2}{v - u}$

$\quad = \dfrac{(v + u)(v - u)}{v - u}$

$\quad = v + u$

**Exam tip:** The formula $y - b = m(x - a)$ will appear in several topics throughout the course. Make sure you are able to use it accurately and consistently.

# Functions and graphs (1)

**Q1** A function is given by $f(x) = \dfrac{5}{x - 2}$. What is a suitable domain of $f$?

**Q2** A function is given by $f(x) = \sqrt{8 - x}$. What is a suitable domain of $f$?

**Q3** Given the graph of $f(x)$, how would you draw the graph of $-f(x)$?

**Q4** Given the graph of $f(x)$, how would you draw the graph of $f(x) + 2$?

**ANSWERS** ▶▶

1  A suitable domain is $x \neq 2$, as the denominator cannot equal 0.

2  As the square root cannot be negative, $8 - x \geq 0$,
   so the domain is $x \leq 8$.

3  You would reflect the graph in the $x$-axis.

4  You would move the graph up two units in the direction of the $y$-axis.

# Functions and graphs (2)

**Q1** Given the graph of $f(x)$, how would you draw the graph of $f(-x)$?

**Q2** Given the graph of $f(x)$, how would you draw the graph of $f(x - 3)$?

**Q3** Given the graph of $f(x)$, how would you draw the graph of $kf(x)$?

**Q4** Given the graph of $f(x)$, how would you draw the graph of $2 - f(x)$?

**ANSWERS** »

**1** You would reflect the graph in the $y$-axis.

**2** You would move the graph three units to the right in the direction of the $x$-axis.

**3** You would stretch it vertically for $k > 1$ or compress it vertically for $k < 1$.

**4** Think of $2 - f(x)$ as $-f(x) + 2$, then reflect the graph in the $x$-axis, after which move it up two units in the direction of the $y$-axis.

# Functions and graphs (3)

**Q1** If $f(x) = 3 - 2x$ and $g(x) = 4x - 3$, find a formula for $f(g(x))$.

**Q2** If $f(x) = 3x + 5$ and $g(x) = 2 + x^2$, find a formula for $g(f(x))$.

**Q3** $f(x) = 2x + 4$ and $g(x) = \frac{1}{2}(x - 4)$.

By finding a formula for $g(f(x))$, state the relationship between $f(x)$ and $g(x)$.

**Q4** Given that $f(x) = 3x + 1$ and $g(x) = x^2 - 1$, find a formula for $f(g(x))$.

**ANSWERS** ⟩⟩

**1** $f(g(x)) = f(4x - 3)$

$\qquad = 3 - 2(4x - 3)$

$\qquad = 3 - 8x + 6 = 9 - 8x$

**2** $g(f(x)) = g(3x + 5)$

$\qquad = 2 + (3x + 5)^2$

$\qquad = 9x^2 + 30x + 27$

**3** $g(f(x)) = g(2x + 4)$

$\qquad = \dfrac{1}{2}(2x + 4 - 4) = x$

Hence $f(x)$ and $g(x)$ are inverse functions.

**4** $f(g(x)) = f(x^2 - 1)$

$\qquad = 3(x^2 - 1) + 1$

$\qquad = 3x^2 - 2$

# Functions and graphs (4)

**Q1** If $f(x) = 3x + 5$ and $g(x) = 2x^2$, evaluate $f(g(-2))$.

**Q2** Given the graph of $f(x)$, how would you draw the graph of $-f(-x)$?

**Q3** If $f(x) = x^2 + 2x$ and $g(x) = x + 12$, find the values of $x$ for which $f(x) = g(x)$.

**Q4** Given the graph of a function, how would you draw the graph of its inverse?

**ANSWERS** ▶▶

**1** $f(g(-2)) = f(2 \times (-2)^2)$

$\qquad = f(8)$

$\qquad = 3 \times 8 + 5$

$\qquad = 29$

**2** You would give $f(x)$ a half-turn (180°) rotation about the origin O.

**3** $f(x) = g(x)$

$\Rightarrow \qquad x^2 + 2x = x + 12$

$\Rightarrow \quad x^2 + x - 12 = 0$

$\Rightarrow \quad (x + 4)(x - 3) = 0$

Hence $x = -4$ or 3.

**4** Reflect the graph in the line $y = x$.

**Exam tip:** Given the graph of a function $f$, you must be able to draw the graph of related functions, for example

$$y = -f(x), y = 3f(x), y = f(x) + 2, y = f'(x), \text{ etc.,}$$

so make sure you know the key points for doing this.

# Differentiation (1)

**Q1** If $y = 4x^3 + 5x^2 - 3x + 2$, find $\dfrac{dy}{dx}$.

**Q2** If $f(x) = 6\sqrt{x}$, find $f'(x)$.

**Q3** If $f(x) = (x + 5)^2$, find $f'(x)$.

**Q4** Find the gradient of the tangent to the curve $y = x^3 - 5x + 1$ at the point where $x = 2$.

**ANSWERS** ▶▶

**1** $\dfrac{dy}{dx} = 12x^2 + 10x - 3$

**2** $f(x) = 6\sqrt{x}$

$= 6x^{1/2}$

$\Rightarrow f'(x) = 3x^{-1/2}$

$= \dfrac{3}{\sqrt{x}}$

**3** $f(x) = (x + 5)^2 = x^2 + 10x + 25$

$\Rightarrow f'(x) = 2x + 10$

**4** $\dfrac{dy}{dx} = 3x^2 - 5$

Hence the gradient of the tangent $= 3 \times 2^2 - 5 = 7$.

# Differentiation (2)

**Q1** If $f(x) = x^3 - 3x - 5$, calculate $f'(-2)$.

**Q2** Find the equation of the tangent to the curve $y = 2x^2 - 3x$ at the point $(2, 2)$.

**Q3** If $y = \dfrac{x^3 - 2x^2 + 5}{x}$, find $\dfrac{dy}{dx}$.

**Q4** If $f(x) = \dfrac{(x + 1)^2}{\sqrt{x}}$, find $f'(x)$.

**ANSWERS** ▶▶

**1**  $f'(x) = 3x^2 - 3$

$\Rightarrow f'(-2) = 3 \times (-2)^2 - 3$

$= 9$

**2**  $\dfrac{dy}{dx} = 4x - 3$

$\Rightarrow m = 4 \times 2 - 3 = 5$ when $x = 2$

Using the formula $y - b = m(x - a)$, the equation of the tangent is
$y - 2 = 5(x - 2)$.

**3**  $y = \dfrac{x^3 - 2x^2 + 5}{x} = \dfrac{x^3}{x} - \dfrac{2x^2}{x} + \dfrac{5}{x}$

$= x^2 - 2x + 5x^{-1}$

$\Rightarrow \qquad \dfrac{dy}{dx} = 2x - 2 - 5x^{-2}$

**4**  $f(x) = \dfrac{(x + 1)^2}{\sqrt{x}}$

$= \dfrac{x^2 + 2x + 1}{x^{1/2}}$

$= \dfrac{x^2}{x^{1/2}} + \dfrac{2x}{x^{1/2}} + \dfrac{1}{x^{1/2}}$

$= x^{3/2} + 2x^{1/2} + x^{-1/2}$

Hence $f'(x) = \dfrac{3}{2}x^{1/2} + x^{-1/2} - \dfrac{1}{2}x^{-3/2}$.

# Differentiation (3)

**Q1** When is a function $f(x)$ said to be *increasing*?

**Q2** Find the interval in which the function $f(x) = 3x^2 + 12x - 5$ is decreasing.

**Q3** Find the stationary points on the curve $y = 2x^3 - 3x^2$.

**Q4** Find the nature of the stationary points in Q3.

**ANSWERS** ⟫

1 When $f'(x) > 0$.

2 $f(x)$ is decreasing when $f'(x) < 0$, that is when $6x + 12 < 0 \Rightarrow x < -2$.

3 Stationary points occur when $\dfrac{dy}{dx} = 0$

$\Rightarrow 6x^2 - 6x = 0$

$\Rightarrow 6x(x - 1) = 0$

that is when $x = 0$ or 1.

By substitution, stationary points are $(0, 0)$ and $(1, -1)$.

4 $(0, 0)$ is a maximum stationary point and $(1, -1)$ is a minimum stationary point.

## Differentiation (4)

**Q1** For which value of $x$ does the function $f(x) = 6x - x^2$ have its greatest value?

**Q2** Find the derivative of $\dfrac{3}{x^2}$.

**Q3** What is the gradient of the curve $y = x^3$ at the point $(-2, -8)$?

**Q4** What is the rate of change of the function $f(x) = 4x^3 - x^2$ at $x = \dfrac{1}{2}$?

**ANSWERS** ▶▶

**1** The greatest value occurs when $f'(x) = 0$

$\Rightarrow 6 - 2x = 0$

$\Rightarrow \quad x = 3.$

**2** As $\dfrac{3}{x^2} = 3x^{-2}$, the derivative is $-6x^{-3}$ or $-\dfrac{6}{x^3}$.

**3** $\dfrac{dy}{dx} = 3x^2$

$\qquad = 3 \times (-2)^2$

$\qquad = 3 \times 4 = 12$ when $x = -2$, so the gradient is 12.

**4** $f'(x) = 12x^2 - 2x$

$\qquad = 12 \times \left(\dfrac{1}{2}\right)^2 - 2 \times \dfrac{1}{2}$

$\qquad = 12 \times \dfrac{1}{4} - 1$

$\qquad = 3 - 1 = 2$

# Differentiation (5)

**Q1** Explain why the function $f(x) = x^3 + 3x^2 + 3x$ never decreases.

**Q2** Find the $x$-coordinate of the point on the curve $y = \dfrac{4}{x^2}$ at which the tangent to the curve has a gradient of 1.

**Q3** The graph of $y = f(x)$ is a parabola. Describe the graph of $y = f'(x)$.

**Q4** The graph of $y = f(x)$ has stationary points at $(3, -5)$ and $(7, 4)$. Write down the coordinates of two points on the graph of $y = f'(x)$.

**ANSWERS** ⟩⟩

**1** $f'(x) = 3x^2 + 6x + 3$

$\qquad = 3(x^2 + 2x + 1)$

$\qquad = 3(x + 1)^2$

As this can never be negative, the function never decreases.

**2** $y = \dfrac{4}{x^2} = 4x^{-2} \quad \Rightarrow \quad \dfrac{dy}{dx} = -8x^{-3}$

$\qquad\qquad\qquad\qquad\qquad = -\dfrac{8}{x^3}.$

Hence $-\dfrac{8}{x^3} = 1$

$\Rightarrow \qquad x^3 = -8$

$\Rightarrow \qquad x = -2.$

**3** It is a straight line.

**4** $(3, 0)$ and $(7, 0)$, because $f'(x) = 0$ at stationary points.

# Differentiation (6)

**Q1** Find where the curve with equation $y = 5x - x^3$ intersects the $x$-axis.

**Q2** At a stationary point on a curve, the gradient changes from negative to zero to positive. What is the nature of the stationary point?

**Q3** The speed $v$ m/s at time $t$ seconds is given by $v(t) = 3t^2 - t$. What is the rate of change of speed when $t = 4$?

**Q4** Find the greatest and least values of $f(x) = x^2$ on the closed interval $-1 \leq x \leq 3$.

**ANSWERS ▶▶**

**1** The curve intersects the x-axis where $5x - x^3 = 0 \Rightarrow x(5 - x^2) = 0$

$\Rightarrow x = 0$ or $\pm\sqrt{5}$.

Therefore, it intersects at $(-\sqrt{5}, 0)$, $(0, 0)$ and $(\sqrt{5}, 0)$.

**2** It is a minimum stationary point.

**3** $\dfrac{dv}{dt} = 6t - 1$

Hence rate of change of speed $= 6 \times 4 - 1 = 23 \, \text{m/s}^2$ when $t = 4$.

**4** At the endpoints, $f(-1) = (-1)^2 = 1$ and $f(3) = 3^2 = 9$.

There is a stationary point when $f'(x) = 0 \Rightarrow 2x = 0 \Rightarrow x = 0$.

As $f(0) = 0$, the greatest value $= 9$ and the least value $= 0$.

**_Exam_ tip** 1: Remember that the derivative $f'(x)$ or $\dfrac{dy}{dx}$ is the gradient of the tangent to a curve.

**_Exam_ tip** 2: If you are asked to find the maximum (greatest) or minimum (least) value of a function $f(x)$, this is an optimisation problem and you should proceed by solving the equation $f'(x) = 0$.

# Recurrence relations (1)

**Q1** A recurrence relation is given by the formula $u_{n+1} = 2u_n + 1$.
If $u_0 = 2$, find $u_1$ and $u_2$.

**Q2** In the recurrence relation $u_{n+1} = mu_n + c$, $u_0 = 4$, $u_1 = 10$ and $u_2 = 13$.
Find $m$ and $c$.

**Q3** Explain why the recurrence relation $u_{n+1} = 0.4u_n + 6$ has a limit.

**Q4** Find the limit of the recurrence relation in Q3.

**ANSWERS** ▶▶

**1**   $u_1 = 2u_0 + 1 = 2 \times 2 + 1 = 5$

    $u_2 = 2u_1 + 1 = 2 \times 5 + 1 = 11$

**2**   $u_1 = mu_0 + c \qquad \Rightarrow 10 = 4m + c$

    $u_2 = mu_1 + c \qquad \Rightarrow 13 = 10m + c$

    Solving simultaneous equations leads to $m = 0.5$ and $c = 8$.

**3**   It has a limit because $-1 < 0.4 < 1$.

**4**   $L = \dfrac{b}{1 - a}$

      $= \dfrac{6}{1 - 0.4}$

      $= \dfrac{6}{0.6} = 10$

# Recurrence relations (2)

**Q1** A recurrence relation is given by the formula $u_{n+1} = 1.1u_n$.
If $u_0 = 50$, calculate $u_3$.

**Q2** Each day 75% of the litter in a street is cleared, but 30 kg is dropped.
Set up a recurrence relation to illustrate this.

**Q3** A sequence is defined by the recurrence relation
$u_{n+1} = 0.2u_n + 5$ with $u_8 = 20$.
Calculate $u_{10}$.

**Q4** A sequence is generated by the recurrence relation
$u_{n+1} = 0.7u_n - 24$.
What is the limit ($L$) of this sequence as $n \rightarrow \infty$?

**ANSWERS** ▶▶

**1**   $u_3 = 1 \cdot 1^3 \times u_0$

$\qquad = 1 \cdot 1^3 \times 50$

$\qquad = 66 \cdot 55$

**2**   $u_{n+1} = 0 \cdot 25 u_n + 30$

**3**   $u_9 = 0 \cdot 2 u_8 + 5 = 0 \cdot 2 \times 20 + 5 = 9$

$\qquad u_{10} = 0 \cdot 2 u_9 + 5 = 0 \cdot 2 \times 9 + 5 = 6 \cdot 8$

**4**   $L = \dfrac{b}{1 - a}$

$\qquad = \dfrac{-24}{1 - 0 \cdot 7}$

$\qquad = \dfrac{-24}{0 \cdot 3}$

$\qquad = -80$

# Recurrence relations (3)

**Q1** If the two sequences $u_{n+1} = 0{\cdot}6u_n + c$ and $u_{n+1} = 0{\cdot}2u_n + c^2$ have the same limit, find $c$.

**Q2** In the recurrence relation $u_{n+1} = 3u_n + 2$, $u_3 = 44$. Find $u_2$.

**Q3** A sequence is defined by the recurrence relation
$u_{n+1} = 0{\cdot}3u_n + 4$, $u_0 = 100$.
What is the smallest value of $n$ for which $u_n < 15$?

**Q4** A sequence is defined by the recurrence relation
$u_{n+1} = 2u_n + 10$.
If $u_m = 3$ what is the value of $u_{m+3}$?

**ANSWERS** ⟩⟩

**1** $L = \dfrac{b}{1 - a}$

$\Rightarrow \quad \dfrac{c}{1 - 0.6} = \dfrac{c^2}{1 - 0.2}$

$\Rightarrow \quad 0.8c = 0.4c^2$

$\Rightarrow \quad \dfrac{c^2}{c} = \dfrac{0.8}{0.4}$

$\Rightarrow \quad c = 2$

**2** $u_{3,} = 3u_2 + 2$

$\Rightarrow \quad 44 = 3u_2 + 2$

$\Rightarrow \quad u_2 = 14$

**3** $u_1 = 0.3u_0 + 4 = 0.3 \times 100 + 4 = 34$

$u_2 = 0.3u_1 + 4 = 0.3 \times 34 + 4 = 14.2$

Hence $n = 1$.

**4** $u_{m+1} = 2u_m + 10 = 16$

$u_{m+2} = 2u_{m+1} + 10 = 42$

$u_{m+3} = 2u_{m+2} + 10 = 94$

# Recurrence relations (4)

**Q1** A sequence is defined by the recurrence relation
$u_{n+1} = (a + 3)u_n + 4$.
For what values of $a$ does this sequence have a limit?

**Q2** If the limit of the recurrence relation $u_{n+1} = 0.75u_n + w$ is 48, find $w$.

**Q3** If the recurrence relation
$u_{n+1} = pu_n + 8$
has a limit of 20 as $n \to \infty$, find $p$.

**Q4** Find the limit of the recurrence relation
$u_{n+1} = -0.1u_n - 22$
as $n \to \infty$.

**ANSWERS** ▶▶

**1** There is a limit if $-1 < a + 3 < 1 \Rightarrow -4 < a < -2$.

**2** $L = \dfrac{b}{1 - a} \Rightarrow \dfrac{w}{1 - 0.75} = 48$

$\Rightarrow w = 48 \times (1 - 0.75) = 12$

**3** $L = \dfrac{b}{1 - a} \Rightarrow \dfrac{8}{1 - p} = 20$

$\Rightarrow \qquad 20(1 - p) = 8$

$\Rightarrow \qquad 20 - 20p = 8$

$\Rightarrow \qquad \qquad 20p = 12$

Hence $\qquad \qquad p = 0.6$.

**4** $L = \dfrac{b}{1 - a}$

$\quad = \dfrac{-22}{1 + 0.1}$

$\quad = \dfrac{-22}{1.1}$

$\quad = \dfrac{-220}{11} = -20$

***Exam* tip:** If you are finding the limit of a recurrence relation

$$u_{n+1} = mu_n + c,$$

remember to justify that the limit exists by writing down that $-1 < m < 1$.

## Identities and radians (1)

**Q1** What is the *exact* value of $\sin 60°$?

**Q2** What is the *exact* value of $\tan \frac{\pi}{6}$?

**Q3** Solve the equation $\sin x° = -\frac{1}{\sqrt{2}}$, $0 \leq x \leq 360$.

**Q4** Solve the equation $\sin x° = \frac{1}{2}$, $0 \leq x \leq 2\pi$.

**ANSWERS** ❱❱

**1** $\dfrac{\sqrt{3}}{2}$

**2** $\tan\dfrac{\pi}{6} = \tan 30° = \dfrac{1}{\sqrt{3}}$

**3** $x = 225$ or $315$

**4** $x = \dfrac{\pi}{6}$ or $\dfrac{5\pi}{6}$

# Identities and radians (2)

**Q1** Convert $135°$ into radians.

**Q2** Convert $\frac{2\pi}{3}$ into degrees.

**Q3** What is the *exact* value of $\tan\frac{7\pi}{4}$?

**Q4** Prove that $\frac{\cos^3 x}{1 - \sin^2 x} = \cos x$.

ANSWERS »

**1**  $135° = \dfrac{135}{180}\pi = \dfrac{3\pi}{4}$ radians

**2**  $\dfrac{2\pi}{3} = \dfrac{2}{3} \times 180° = 120°$

**3**  $\tan \dfrac{7\pi}{4} = \tan \left( \dfrac{7}{4} \times 180 \right)°$

$= \tan 315°$

$= -\tan 45°$

$= -1$

**4**  Left side $= \dfrac{\cos^3 x}{1 - \sin^2 x}$

$= \dfrac{\cos^3 x}{\cos^2 x + \sin^2 x - \sin^2 x}$

$= \dfrac{\cos^3 x}{\cos^2 x}$

$= \cos x = $ Right side

## Identities and radians (3)

**Q1** What is the least period of $4\tan 3x°$?

**Q2** Solve the equation $2\sin x° = \sqrt{3}$, $0 \le x \le 360$.

**Q3** State the coordinates of the maximum turning point of the graph of $y = \cos(x − 45)°$, $0 \le x \le 360$.

**Q4** Solve the equation $\sin 2x° = 0$, $0 \le x \le 360$.

ANSWERS ⟩⟩

**1** $180° \div 3 = 60°$

**2** $2\sin x° = \sqrt{3}$

$\Rightarrow \quad \sin x° = \dfrac{\sqrt{3}}{2}$

$\Rightarrow \qquad x = 60 \text{ or } 120$

**3** Coordinates are (45, 1).

**4** $2x = 0, 180, 360, 540, 720$

$\Rightarrow x = 0, 90, 180, 270, 360$

# Identities and radians (4)

**Q1** Simplify $\cos(-x)^\circ$.

**Q2** Solve the equation $2\cos 3y^\circ = 1$, $\quad 0 \le x \le 360$.

**Q3** Prove that $2\cos^2 A + 3\sin^2 A - 2 = \sin^2 A$.

**Q4** Solve the equation $2\cos\left(2x + \dfrac{\pi}{4}\right) = 1$, $\quad 0 \le x \le 2\pi$.

**ANSWERS** 〉〉

**1** $\cos x°$

**2** $2\cos 3y° = 1 \Rightarrow \cos 3y° = \frac{1}{2} \Rightarrow 3y = 60, 300, 420, 660, 780, 1020$

Hence $y = 20, 100, 140, 220, 260, 340$.

**3** Left side $= 2\cos^2 A + 3\sin^2 A - 2$

$$= 2(1 - \sin^2 A) + 3\sin^2 A - 2$$

$$= 2 - 2\sin^2 A + 3\sin^2 A - 2$$

$$= \sin^2 A = \text{Right side}$$

**4** $2\cos\left(2x + \frac{\pi}{4}\right) = 1$

$\Rightarrow \quad \cos\left(2x + \frac{\pi}{4}\right) = \frac{1}{2}$

$\Rightarrow \quad 2x + \frac{\pi}{4} = \frac{\pi}{3}, \frac{5\pi}{3}, \frac{7\pi}{3}, \frac{11\pi}{3}$

Hence $\quad 2x = \frac{\pi}{12}, \frac{17\pi}{12}, \frac{25\pi}{12}, \frac{41\pi}{12}$

$\Rightarrow \quad x = \frac{\pi}{24}, \frac{17\pi}{24}, \frac{25\pi}{24}, \frac{41\pi}{24}$

***Exam* tip:** Remember that $180° = \pi$ radians.

## Quadratic theory (1)

**Q1** Solve the equation $20x - 5x^2 = 0$.

**Q2** Solve the equation $3x^2 - 10x - 8 = 0$.

**Q3** Solve the equation $2x^2 - 50 = 0$.

**Q4** What is the quadratic formula for solving the equation $ax^2 + bx + c = 0$?

**ANSWERS** ))

**1** $20x - 5x^2 = 0$

$\Rightarrow 5x(4 - x) = 0$

$\Rightarrow \qquad x = 0 \text{ or } 4$

**2** $3x^2 - 10x - 8 = 0$

$\Rightarrow (3x + 2)(x - 4) = 0$

$\Rightarrow \qquad x = -\dfrac{2}{3} \text{ or } 4$

**3** $2x^2 - 50 = 0$

$\Rightarrow \qquad 2(x^2 - 25) = 0$

$\Rightarrow 2(x + 5)(x - 5) = 0$

$\Rightarrow \qquad x = -5 \text{ or } 5$

**4** $x = \dfrac{-b \pm \sqrt{b^2 - 4ac}}{2a}$

# Quadratic theory (2)

**Q1** When does the quadratic equation $ax^2 + bx + c = 0$ have equal roots?

**Q2** What type of roots does the equation $x^2 + 5x - 4 = 0$ have?

**Q3** What can you say about the roots of the equation $2x^2 + 8x + 6 = 0$?

**Q4** Express $x^2 + 4x + 5$ in the form $(x + a)^2 + b$.

**ANSWERS** 》

**1** The equation has equal roots when $b^2 - 4ac = 0$.

**2** As $b^2 - 4ac = 5^2 - 4 \times 1 \times (-4) = 41$, which is positive, the roots are real.

**3** $b^2 - 4ac = 8^2 - 4 \times 2 \times 6 = 16$

As this is positive and a perfect square, the roots are real and rational.

**4** $x^2 + 4x + 5 = x^2 + 4x + 4 + 1$

$$= (x + 2)^2 + 1$$

# Quadratic theory (3)

**Q1** Express $2x^2 + 4x + 5$ in the form $2(x + p)^2 + q$.

**Q2** For what values of $h$ does the equation $5x^2 + 10x + h = 0$ have no real roots?

**Q3** Is the turning point of the parabola $y = 2x^2 + 4x - 5$ a maximum or a minimum?

**Q4** Solve the inequality $x^2 + 2x - 8 < 0$.

**ANSWERS ▶▶**

**1**  $2x^2 + 4x + 5 = 2(x^2 + 2x) + 5$

$= 2(x^2 + 2x + 1) - 2 + 5$

$= 2(x + 1)^2 + 3$

**2**  It has no real roots when $b^2 - 4ac < 0 \Rightarrow 10^2 - 4 \times 5 \times h < 0$

$\Rightarrow 100 - 20h < 0$, that is when $h > 5$.

**3**  As the coefficient of the $x^2$ term, 2, $> 0$, the turning point is a minimum.

**4**  By factorising $x^2 + 2x - 8 = (x + 4)(x - 2)$ and sketching the parabola $y = x^2 + 2x - 8$, the solution is $-4 < x < 2$.

# Quadratic theory (4)

**Q1** Expand the brackets and simplify $(\sqrt{2} + \sqrt{6})^2$.

**Q2** Write down the coordinates of the turning point of the parabola with equation $y = 8 - (x + 5)^2$.

**Q3** Write down a quadratic equation with roots of 5 and $-3$.

**Q4** Find the values of $k$ for which the equation $x^2 + 2kx + 9 = 0$ has equal roots.

**ANSWERS** ❯❯

**1** $(\sqrt{2} + \sqrt{6})^2 = (\sqrt{2} + \sqrt{6})(\sqrt{2} + \sqrt{6})$

$$= 2 + 2\sqrt{12} + 6$$

$$= 8 + 2\sqrt{4 \times 3}$$

$$= 8 + 4\sqrt{3}$$

**2** There is a maximum turning point at $(-5, 8)$.

**3** $(x - 5)(x + 3) = 0$

$\Rightarrow \quad x^2 - 2x - 15 = 0$

**4** There are equal roots when

$b^2 - 4ac = 0$

$\Rightarrow (2k)^2 - 4 \times 1 \times 9 = 0$

$\Rightarrow \qquad\qquad 4k^2 = 36.$

Hence $k^2 = 9 \Rightarrow k = -3$ or 3.

## Quadratic theory (5)

**Q1** By expressing in the form $y = (x + a)^2 + b$, find the coordinates of the turning point of the parabola with equation $y = x^2 + 6x - 1$.

**Q2** Check your answer in Q1 using calculus.

**Q3** State the nature of the turning point in Q1.

**Q4** For what values of $x$ is $y$ undefined if $y = \dfrac{3}{x^2 + 2x - 15}$ ?

**ANSWERS ▶▶**

**1** $y = x^2 + 6x - 1 = x^2 + 6x + 9 - 10 = (x + 3)^2 - 10$

Hence the turning point is $(-3, -10)$.

**2** $y = x^2 + 6x - 1$

$\Rightarrow \dfrac{dy}{dx} = 2x + 6 = 0$ at the turning point.

Hence $x = -3$, which leads to $(-3, -10)$.

**3** It is a minimum turning point.

**4** $y$ is undefined when $x^2 + 2x - 15 = 0$

$\Rightarrow \quad (x + 5)(x - 3) = 0$

$\Rightarrow \qquad\qquad\qquad x = -5 \text{ or } 3.$

***Exam* tip:** Make sure you know how the discriminant $b^2 - 4ac$ determines whether the roots of a quadratic equation are real, not real or equal.

# Polynomials (1)

**Q1** Evaluate $f(-2)$ where $f(x) = x^3 - 5x + 3$.

**Q2** Find the remainder on dividing $2x^3 - x^2 + 3x - 8$ by $(x - 1)$.

**Q3** Show that $(x - 2)$ is a factor of $x^3 + x^2 - 12$.

**Q4** Factorise $x^3 - x^2 - 4x + 4$.

**ANSWERS** ▶▶

**1**  $f(-2) = (-2)^3 - 5 \times (-2) + 3$

$\qquad = -8 + 10 + 3 = 5$

**2**

$$
\begin{array}{r|rrrr}
1 & 2 & -1 & 3 & -8 \\
  &   & 2 & 1 & 4 \\
\hline
  & 2 & 1 & 4 & -4
\end{array}
$$

Remainder = $-4$.

**3**

$$
\begin{array}{r|rrrr}
2 & 1 & 1 & 0 & -12 \\
  &   & 2 & 6 & 12 \\
\hline
  & 1 & 3 & 6 & 0
\end{array}
$$

$(x - 2)$ is a factor as the remainder = 0.

**4**

$$
\begin{array}{r|rrrr}
1 & 1 & -1 & -4 & 4 \\
  &   & 1 & 0 & -4 \\
\hline
  & 1 & 0 & -4 & 0
\end{array}
$$

Hence $x^3 - x^2 - 4x + 4 = (x - 1)(x^2 - 4)$

$\qquad\qquad\qquad\qquad\quad = (x - 1)(x - 2)(x + 2).$

## Polynomials (2)

**Q1** Solve the equation $x^3 - x^2 - 4x + 4 = 0$ (see Q4 in Polynomials (1)).

**Q2** Show that the equation $x^3 - 2x^2 + 4x - 5 = 0$ has a root between 1 and 2.

**Q3** A cubic curve crosses the x-axis at 1, 2 and 3 and the y-axis at $-12$. Find the equation of the curve.

**Q4** What is the remainder when $4x^2 + 6x - 6$ is divided by $2x - 1$?

**ANSWERS** 》》

**1** $x = 1, 2$ or $-2$

**2** Let $f(x) = x^3 - 2x^2 + 4x - 5$

Then $f(1) = -2$ (negative) and $f(2) = 3$ (positive).

Hence there is a root between 1 and 2.

**3** The equation is $y = k(x - 1)(x - 2)(x - 3)$.

Substitute $(0, -12)$ into the equation to find $k = 2$.

Hence equation is $y = 2(x - 1)(x - 2)(x - 3)$.

**4** $\dfrac{1}{2}$

| | 4 | 6 | $-6$ |
|---|---|---|---|
| | | 2 | 4 |
| | 4 | 8 | $-2$ |

Remainder $= -2$.

# Polynomials (3)

**Q1** If $x - 3$ is a factor of $6x^3 - 25x^2 + kx - 60$, find $k$.

**Q2** What is the remainder when $3x^3 - 7x^2 + 6x - 8$ is divided by $3x + 2$?

**Q3** If $x^3 - 3x^2 + 2x + 1 = (x - 3)(x^2 + 2) + 2p$, find $p$.

**Q4** Prove that $x - 3$ is a factor of $x^3 - 2x^2 - 5x + 6$.

**ANSWERS** ▶▶

**1** 

$$\begin{array}{r|rrrr} 3 & 6 & -25 & k & -60 \\ & & 18 & -21 & 3k - 63 \\ \hline & 6 & -7 & k - 21 & 3k - 123 \end{array}$$

Hence $3k - 123 = 0$ and $k = 41$.

**2**

$$\begin{array}{r|rrrr} -\dfrac{2}{3} & 3 & -7 & 6 & -8 \\ & & -2 & 6 & -8 \\ \hline & 3 & -9 & 12 & -16 \end{array}$$

Remainder $= -16$.

**3** $x^3 - 3x^2 + 2x + 1 = (x - 3)(x^2 + 2) + 2p$

$$= x^3 - 3x^2 + 2x - 6 + 2p.$$

Hence $-6 + 2p = 1$, so $p = 3 \cdot 5$.

**4**

$$\begin{array}{r|rrrr} 3 & 1 & -2 & -5 & 6 \\ & & 3 & 3 & -6 \\ \hline & 1 & 1 & -2 & 0 \end{array}$$

$x - 3$ is a factor because the remainder is 0.

# Polynomials (4)

**Q1** Find the value of $k$ if $kx^4 - 4x^2 + 3x - 2$ has remainder of 115 on division by $x + 3$.

**Q2** Show that $x = 1$ is the only root of the equation $x^3 - 5x^2 + 9x - 5 = 0$.

**Q3** For what value of $f$ is $x - 3$ a factor of $x^3 - 2x^2 - 5x + f$?

**Q4** Find the coordinates of the points where the line $y = 5x + 3$ intersects the curve $y = x^3 - x^2$.

**ANSWERS** ▶▶

**1**

$$\begin{array}{r|rrrrr} -3 & k & 0 & -4 & 3 & -2 \\ & & -6 & 18 & -42 & 117 \\ \hline & 2 & -6 & 14 & -39 & 115 \end{array}$$

Hence $k = 2$. Hint: Work backwards from 115.

**2**

$$\begin{array}{r|rrrr} 1 & 1 & -5 & 9 & -5 \\ & & 1 & -4 & 5 \\ \hline & 1 & -4 & 5 & 0 \end{array}$$

Hence $x^3 - 5x^2 + 9x - 5 = (x - 1)(x^2 - 4x + 5)$ and as $x^2 - 4x + 5 = 0$ has no real roots because $b^2 - 4ac = -4$, the only root is $x = 1$.

**3**

$$\begin{array}{r|rrrr} 3 & 1 & -2 & -5 & f \\ & & 3 & 3 & -6 \\ \hline & 1 & 1 & -2 & 0 \end{array}$$

Hence $f = 6$.

**4** The line intersects the curve where

$$x^3 - x^2 = 5x + 3 \Rightarrow x^3 - x^2 - 5x - 3 = 0.$$

$$\begin{array}{r|rrrr} -1 & 1 & -1 & -5 & -3 \\ & & -1 & 2 & 3 \\ \hline & 1 & -2 & -3 & 0 \end{array}$$

Hence $x^3 - x^2 - 5x - 3 = (x + 1)(x^2 - 2x - 3) = 0$,

that is $(x + 1)(x + 1)(x - 3) = 0 \Rightarrow x = -1$ or 3,

and the points of intersection are $(-1, -2)$ and $(3, 18)$.

***Exam* tip:** The method of synthetic division is extremely important in helping you to factorise polynomials and solve problems that involve polynomials, so make sure you are expert at using this method accurately.

# Integration (1)

**Q1** Find $\int (3x^2 + 2x - 5)\,dx$.

**Q2** Integrate $x\sqrt{x}$ with respect to $x$.

**Q3** $f'(x) = 6x^2 - 4x + 3$. Find $f(x)$ if $f(1) = 5$.

**Q4** $\dfrac{dy}{dx} = 8x - 3$. If $y = 7$ when $x = 2$, find an equation for $y$.

ANSWERS ❯❯

**1** $\int(3x^2 + 2x - 5)dx = x^3 + x^2 - 5x + C$

**2** $\int(x\sqrt{x})dx = \int(x \times x^{1/2})dx$

$$= \int x^{3/2}dx$$

$$= \frac{2}{5}x^{5/2} + C$$

**3** $f(x) = \int(6x^2 - 4x + 3)dx$

$$= 2x^3 - 2x^2 + 3x + C$$

$f(1) = 5$

$\Rightarrow 5 = 2 - 2 + 3 + C$

$\Rightarrow C = 2$

Hence $f(x) = 2x^3 - 2x^2 + 3x + 2.$

**4** $y = \int(8x - 3)dx$

$$= 4x^2 - 3x + C$$

If $y = 7$ when $x = 2$,

$7 = 4 \times 2^2 - 3 \times 2 + C.$

Hence $C = -3$ and $y = 4x^2 - 3x - 3.$

## Integration (2)

**Q1** Evaluate $\int_1^3 (2x - 4)\,dx$.

**Q2** Evaluate $\int_0^4 \sqrt{x}\,dx$.

**Q3** Calculate the area under the curve $y = x^2$ from $x = 3$ to $x = 6$.

**Q4** What does it mean if the area defined by $\int_a^b f(x)\,dx$ is negative?

ANSWERS ▶▶

**1** $\int\limits_{1}^{3} (2x - 4)\,dx = \left[x^2 - 4x\right]_{1}^{3}$

$$= (3^2 - 4 \times 3) - (1^2 - 4 \times 1)$$

$$= (-3) - (-3) = 0$$

**2** $\int\limits_{0}^{4} \sqrt{x}\,dx = \int\limits_{0}^{4} x^{1/2}\,dx$

$$= \left[\frac{2}{3}x^{3/2}\right]_{0}^{4}$$

$$= \left[\frac{2}{3}\sqrt{x^3}\right]_{0}^{4}$$

$$= \frac{2}{3}\left(\sqrt{4^3} - \sqrt{0^3}\right)$$

$$= \frac{2}{3} \times 8$$

$$= \frac{16}{3}$$

**3** Area $= \int\limits_{3}^{6} x^2\,dx = \left[\frac{1}{3}x^3\right]_{3}^{6}$

$$= \left(\frac{1}{3} \times 6^3\right) - \left(\frac{1}{3} \times 3^3\right)$$

$$= 72 - 9 = 63$$

**4** It means that the area is *below* the x-axis.

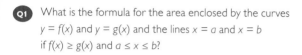

# Integration (3)

**Q1** What is the formula for the area enclosed by the curves
$y = f(x)$ and $y = g(x)$ and the lines $x = a$ and $x = b$
if $f(x) \geq g(x)$ and $a \leq x \leq b$?

**Q2** If you are finding the area enclosed by the curves $y = 4x - x^2$
and $y = x^2 - 2x$, how would you find the limits of integration?

**Q3** For the data in Q2, find the limits.

**Q4** With the limits found in Q3 calculate the area enclosed by
$y = 4x - x^2$ and $y = x^2 - 2x$.

ANSWERS ▶▶

**1** $A = \int_a^b [f(x) - g(x)] \mathrm{d}x$

**2** Solve the equation $4x - x^2 = x^2 - 2x$.

**3** $4x - x^2 = x^2 - 2x$

$\Rightarrow \quad 2x^2 - 6x = 0$

$\Rightarrow \quad 2x(x - 3) = 0$

$\Rightarrow \qquad\qquad x = 0 \text{ or } 3$

**4** Area $= \int_0^3 [(4x - x^2) - (x^2 - 2x)] \mathrm{d}x$

$= \int_0^3 (6x - 2x^2) \mathrm{d}x$

$= \left[ 3x^2 - \frac{2}{3}x^3 \right]_0^3$

$= 27 - 18 = 9$

## Integration (4)

**Q1** Find $\int \left( \dfrac{x^4 - 2x^2 - 5}{x^2} \right) dx$.

**Q2** If $\int_1^a 5 \, dx = 10$, find $a$.

**Q3** Find $\int \left( \dfrac{2}{3\sqrt{x}} \right) dx$.

**Q4** If $f'(x) = 5x^{\frac{1}{4}}$ and $f(1) = 1$, find $f(x)$.

ANSWERS ▶▶

**1** $\int\left(\dfrac{x^4 - 2x^2 - 5}{x^2}\right)dx = \int(x^2 - 2 - 5x^{-2})dx$

$$= \frac{1}{3}x^3 - 2x + 5x^{-1} + C$$

**2** $\int_1^a 5\,dx = [5x]_1^a = 5a - 5 = 10 \Rightarrow a = 3$

**3** $\int\left(\dfrac{2}{3\sqrt{x}}\right)dx = \int\left(\dfrac{2}{3x^{1/2}}\right)dx$

$$= \int\left(\frac{2}{3}x^{-1/2}\right)dx$$

$$= \frac{4}{3}x^{1/2} + C$$

$$= \frac{4}{3}\sqrt{x} + C$$

**4** $f(x) = \int 5x^{1/2}\,dx = 2x^{5/2} + C$

As $f(1) = 1$,    $2 \times 1^{5/2} + C = 1$

$\Rightarrow$            $2 + C = 1$

Hence          $C = -1$ and $f(x) = 2x^{5/2} - 1$.

## Integration (5)

**Q1** Find $\int(x - 3)(x + 2)dx$.

**Q2** If $\dfrac{dy}{dx} = 4x + 2$ and $y = 10$ when $x = 2$, express $y$ in terms of $x$.

**Q3** Find $f(x)$ if $f'(x) = \dfrac{2 - 5x}{x\sqrt{x}}$.

**Q4** Evaluate $\displaystyle\int_{2}^{6} du$.

**ANSWERS** ▶▶

**1** $\int (x - 3)(x + 2)\,dx = \int (x^2 - x - 6)\,dx$

$$= \frac{1}{3}x^3 - \frac{1}{2}x^2 - 6x + C$$

**2** $y = \int (4x + 2)\,dx = 2x^2 + 2x + C$

$$\Rightarrow 10 = 2 \times 2^2 + 2 \times 2 + C \Rightarrow C = -2$$

Hence $y = 2x^2 + 2x - 2$

**3** $f(x) = \int \left( \dfrac{2 - 5x}{x\sqrt{x}} \right) dx = \int \left( \dfrac{2}{x^{3/2}} - \dfrac{5x}{x^{3/2}} \right) dx$

$$= \int (2x^{-3/2} - 5x^{-1/2})\,dx$$

$$= -4x^{-1/2} - 10x^{1/2} + C$$

**4** $\displaystyle\int_{2}^{6} du = \int_{2}^{6} 1\,du = [u]_{2}^{6} = 6 - 2 = 4$

***Exam* tip:** If you meet a question in your exam that features a graph with a shaded area, it is likely that the solution will be found through integration.

# Compound and multiple angles (1)

**Q1** Write down a formula for $\tan A$.

**Q2** What is the value of $\cos^2 A + \sin^2 A$ for all values of $A$?

**Q3** If $\cos a° = \dfrac{4}{5}$ and $0 \le a \le 90$, what is the value of $\sin a°$?

**Q4** If $\cos x° = \dfrac{2}{\sqrt{5}}$ and $0 \le x \le 90$, find the value of $\tan x°$.

**ANSWERS** 〉〉

**1** $\tan A = \dfrac{\sin A}{\cos A}$

**2** $\cos^2 A + \sin^2 A = 1$ for all values of $A$.

**3** Use Pythagoras' theorem and right-angled trigonometry to find $\sin a° = \dfrac{3}{5}$.

**4** Use the above method to find $\tan x° = \dfrac{1}{2}$.

# Compound and multiple angles (2)

**Q1** If $a$ and $b$ are acute angles such that $\sin a° = \frac{4}{5}$ and $\sin b° = \frac{5}{13}$, find the value of $\sin (a + b)°$.

**Q2** For the data in Q1, find the value of $\sin (a - b)°$.

**Q3** For the data in Q1, find the value of $\cos (a + b)°$.

**Q4** For the data in Q1, find the value of $\cos (a - b)°$.

**ANSWERS ▸▸**

**1**  $\sin a° = \dfrac{4}{5} \Rightarrow \cos a° = \dfrac{3}{5}$

*and* $\sin b° = \dfrac{5}{13} \Rightarrow \cos b° = \dfrac{12}{13}$ (Pythagoras)

So $\sin(a + b)° = \sin a° \cos b° + \cos a° \sin b°$

$$= \dfrac{4}{5} \times \dfrac{12}{13} + \dfrac{3}{5} \times \dfrac{5}{13} = \dfrac{48}{65} + \dfrac{15}{65} = \dfrac{63}{65}$$

**2**  $\sin(a - b)° = \sin a° \cos b° - \cos a° \sin b° = \dfrac{48}{65} - \dfrac{15}{65} = \dfrac{33}{65}$

**3**  $\cos(a + b)° = \cos a° \cos b° - \sin a° \sin b° = \dfrac{3}{5} \times \dfrac{12}{13} - \dfrac{4}{5} \times \dfrac{5}{13}$

$$= \dfrac{36}{65} - \dfrac{20}{65} = \dfrac{16}{65}$$

**4**  $\cos(a - b)° = \cos a° \cos b° + \sin a° \sin b° = \dfrac{36}{65} + \dfrac{20}{65} = \dfrac{56}{65}$

# Compound and multiple angles (3)

**Q1** If $\tan x° = \dfrac{8}{6}$, $0 \le x \le 90$, find the value of $\sin 2x°$.

**Q2** If $\sin a° = \dfrac{1}{3}$, $0 \le a \le 90$, find the value of $\sin 2a°$.

**Q3** If $\sin b° = \dfrac{1}{2}$, $0 \le b \le 90$, find the value of $\cos 2b°$.

**Q4** If $\cos y° = \dfrac{3}{4}$, $0 \le y \le 90$, find the value of $\cos 2y°$.

**ANSWERS ▶▶**

**1**  $\tan x° = \dfrac{8}{6} \Rightarrow \sin x° = \dfrac{8}{10}$ and $\cos x° = \dfrac{6}{10}$ (Pythagoras)

So $\sin 2x° = 2 \sin x° \cos x° = 2 × \dfrac{8}{10} × \dfrac{6}{10} = \dfrac{96}{100} = \dfrac{24}{25}$

**2**  $\sin a° = \dfrac{1}{3} \Rightarrow \cos a° = \dfrac{\sqrt{8}}{3}$ (Pythagoras)

So $\sin 2a° = 2 \sin a° \cos a° = 2 × \dfrac{1}{3} × \dfrac{\sqrt{8}}{3} = \dfrac{2\sqrt{8}}{9} = \dfrac{4\sqrt{2}}{9}$

**3**  $\begin{aligned} \cos 2b° &= 1 - 2\sin^2 b° \\ &= 1 - 2 × \left(\dfrac{1}{2}\right)^2 \\ &= 1 - 2 × \dfrac{1}{4} = \dfrac{1}{2} \end{aligned}$

**4**  $\begin{aligned} \cos 2y° &= 2\cos^2 y° - 1 \\ &= 2 × \left(\dfrac{3}{4}\right)^2 - 1 \\ &= 2 × \dfrac{9}{16} - 1 = \dfrac{2}{16} = \dfrac{1}{8} \end{aligned}$

## Compound and multiple angles (4)

**Q1** Change the subject of the formula $\cos 2A = 2\cos^2 A - 1$ to $\cos^2 A$.

**Q2** Change the subject of the formula $\cos 2A = 1 - 2\sin^2 A$ to $\sin^2 A$.

**Q3** By expressing $75°$ as $(45 + 30)°$, find the *exact* value of $\sin 75°$.

**Q4** Prove that $\sin(90° - A) = \cos A$.

**ANSWERS** ▶▶

**1** $\cos^2 A = \dfrac{1 + \cos 2A}{2}$ (This is useful for integrating $\cos^2 A$.)

**2** $\sin^2 A = \dfrac{1 - \cos 2A}{2}$ (This is useful for integrating $\sin^2 A$.)

**3** $\sin 75° = \sin(45 + 30)° = \sin 45° \cos 30° + \cos 45° \sin 30°$

$$= \frac{1}{\sqrt{2}} \times \frac{\sqrt{3}}{2} + \frac{1}{\sqrt{2}} \times \frac{1}{2} = \frac{\sqrt{3} + 1}{2\sqrt{2}}$$

**4** Left side $= \sin(90° - A) = \sin 90° \cos A - \cos 90° \sin A$

$$= 1 \times \cos A - 0 \times \sin A$$

$$= \cos A = \text{Right side}$$

## Compound and multiple angles (5)

**Q1** How would you *start* to solve the equation
$\cos 2x° + 3\sin x° - 2 = 0$?

**Q2** Factorise $2\sin^2 x° - 3\sin x° + 1$.

**Q3** Hence solve the equation $2\sin^2 x° - 3\sin x° + 1 = 0$, $\quad 0 \leq x \leq 360$.

**Q4** Now solve the equation $2\sin^2 x - 3\sin x + 1 = 0$, $\quad 0 \leq x \leq 2\pi$.

**ANSWERS** ▶▶

**1** Replace $\cos 2x°$ by $1 - 2\sin^2 x°$.

**2** $(2\sin x° - 1)(\sin x° - 1)$

**3** $\sin x° = \dfrac{1}{2}$ or $1 \Rightarrow x = 30, 90, 150$

**4** $x = \dfrac{\pi}{6}, \dfrac{\pi}{2}, \dfrac{5\pi}{6}$

**_Exam_ tip:** If you meet a question that involves $\sin(a \pm b)$, $\cos(a \pm b)$, $\sin 2a$ or $\cos 2a$, look up the formulae list before continuing.

# The circle (1)

**Q1** What is the equation of a circle with its centre at the origin and a radius of 5 units?

**Q2** What is the equation of a circle with its centre at $(3, -2)$ and a radius of 4 units?

**Q3** Find the centre of the circle with equation
$x^2 + y^2 + 12x - 6y + 36 = 0$.

**Q4** Find the radius of the circle in Q3.

**ANSWERS**

**1** $x^2 + y^2 = 25$

**2** $(x - 3)^2 + (y + 2)^2 = 16$

**3** $2g = 12 \quad \Rightarrow \quad g = 6$

$\Rightarrow \qquad\qquad -g = -6$

$2f = -6 \quad \Rightarrow \quad f = -3$

$\Rightarrow \qquad\qquad -f = 3$

Hence centre is at $(-6, 3)$.

**4** $r = \sqrt{g^2 + f^2 - c} = \sqrt{36 + 9 - 36} = 3$

# The circle (2)

**Q1** Find the equation of the circle with centre $(3, 1)$ that passes through the point $(5, 7)$.

**Q2** Show that the point $(7, 2)$ lies on the circle
$x^2 + y^2 - 6x + 4y - 19 = 0$.

**Q3** Find the equation of the tangent at the point $(7, 2)$ to the circle in Q2.

**Q4** Given their equations, how would you prove that two circles touch externally?

**ANSWERS ▶▶**

**1** Using the distance formula, $r^2 = (5 - 3)^2 + (7 - 1)^2 = 40$.

Hence the equation is $(x - 3)^2 + (y - 1)^2 = 40$.

**2** By substitution,

$$x^2 + y^2 - 6x + 4y - 19 = 7^2 + 2^2 - 6 \times 7 + 4 \times 2 - 19 = 0.$$

Hence $(7, 2)$ lies on the circle.

**3** The centre of circle is $(3, -2)$.

$$m_{radius} = \frac{2 + 2}{7 - 3} = 1, \text{ so } m_{tangent} = -1.$$

Hence the equation of the tangent is $y - 2 = -1(x - 7)$.

**4** Show that the distance between their centres equals the sum of their radii.

# The circle (3)

**Q1** How would you *start* to find the points of intersection of the line $y = x + 1$ and the circle $x^2 + y^2 + 2x + 4y - 20 = 0$?

**Q2** Show that the line $y = 3$ is a tangent to the circle $x^2 + y^2 - 4x - 2y + 1 = 0$.

**Q3** Find the point of contact in Q2.

**Q4** What is the centre and radius of the circle $(x - 4)^2 + (y + 2)^2 = 25$?

**ANSWERS** ▶▶

**1** Replace $y$ by $(x + 1)$ in the equation of the circle, that is

$x^2 + (x + 1)^2 + 2x + 4(x + 1) - 20 = 0$.

**2** Replace $y$ by 3 in the equation of the circle, that is

$x^2 + 9 - 4x - 6 + 1 = 0$.

Hence $x^2 - 4x + 4 = 0$.

As $b^2 - 4ac = 0$, the line is a tangent to the circle.

**3** $x^2 - 4x + 4 = 0 \Rightarrow (x - 2)^2 = 0 \Rightarrow x = 2$, and hence the point of contact is $(2, 3)$.

**4** The centre is $(4, -2)$ and the radius is $\sqrt{25} = 5$.

# The circle (4)

**Q1** What is the equation of the circle that has the line joining A (4, 3) to B (8, 3) as the diameter?

**Q2** What is the centre of the circle of equation $2x^2 + 2y^2 + 8x - 3 = 0$?

**Q3** Given that the point $(a\sin\theta, a\cos\theta)$ lies on the circle $x^2 + y^2 = 25$, find $a$.

**Q4** A circle passes through the origin, A (0, 6) and B (8, 0). Find its equation.

**ANSWERS** 》》

**1** The centre is the mid-point of AB, that is (6, 3).

As AB = 4 units, the radius = 2 units.

Hence the equation is $(x - 6)^2 + (y - 3)^2 = 4$.

**2** $(-2, 0)$ (Remember to divide by 2.)

**3** $x^2 + y^2 = 25$

$\Rightarrow \quad a^2 \sin^2\theta + a^2 \cos^2\theta = 25$

$\Rightarrow \quad a^2(\sin^2\theta + \cos^2\theta) = 25$

Hence $a^2 = 25 \Rightarrow a = \pm 5$.

**4** As $\angle AOB = 90°$, AB is a diameter.

The centre is the mid-point of AB, that is (4, 3).

AB = 10 units (Pythagoras), so the radius = 5 units.

Hence the equation is $(x - 4)^2 + (y - 3)^2 = 25$.

## The circle (5)

**Q1** Show that the line $y = x$ does not intersect the circle with equation $x^2 + y^2 - 10x - 4y + 25 = 0$.

**Q2** The lines $x = 3$, $x = 11$, $y = -2$, $y = 6$ are tangents to a circle. What is the equation of the circle?

**Q3** Find the radius of the circle $x^2 + y^2 + 12x - 5y = 0$.

**Q4** A circle has the equation $x^2 + y^2 - 4x - 8y + 3 = 0$. Calculate the length of the chord of this circle that is part of the $x$-axis.

**ANSWERS** ▶▶

**1** Replace $y$ by $x$ in the equation of the circle, that is

$x^2 + x^2 - 10x - 4x + 25 = 0$.

Hence $2x^2 - 14x + 25 = 0$.

As $b^2 - 4ac = (-14)^2 - 4 \times 2 \times 25 = -4 < 0$, the line does not intersect the circle.

**2** By sketching the lines, find that the centre is $(7, 2)$ and the radius is 4

Hence the equation is $(x - 7)^2 + (y - 2)^2 = 16$.

**3** $r = \sqrt{g^2 + f^2 - c} = \sqrt{6^2 + (-2.5)^2} = \sqrt{42.25} = 6.5$

**4** The circle cuts the $x$-axis when $y = 0$, that is

$x^2 - 4x + 3 = 0 \Rightarrow (x - 1)(x - 3) = 0$.

Hence $x = 1$ or 3, so the chord length is 2.

***Exam* tip:** If you are asked to find the equation of a circle, calculate the coordinates of the centre and the radius, then use $(x - a)^2 + (y - b)^2 = r^2$.

## Vectors (1)

**Q1** If $\mathbf{a} = \begin{pmatrix} 3 \\ -2 \\ 5 \end{pmatrix}$ and $\mathbf{b} = \begin{pmatrix} -4 \\ -5 \\ 0 \end{pmatrix}$, find the components of $3\mathbf{a} + \mathbf{b}$.

**Q2** A is the point $(4, -2)$ and B is $(5, 2)$.

Find the components of vector $\overrightarrow{AB}$.

**Q3** Find the magnitude of the vector with components $\begin{pmatrix} 3 \\ -4 \\ 12 \end{pmatrix}$.

**Q4** Find the magnitude of the vector $2\mathbf{i} - 3\mathbf{j} + \mathbf{k}$.

**ANSWERS »**

**1** $3\mathbf{a} + \mathbf{b} = 3\begin{pmatrix} 3 \\ -2 \\ 5 \end{pmatrix} + \begin{pmatrix} -4 \\ -5 \\ 0 \end{pmatrix}$

$= \begin{pmatrix} 9 \\ -6 \\ 15 \end{pmatrix} + \begin{pmatrix} -4 \\ -5 \\ 0 \end{pmatrix}$

$= \begin{pmatrix} 5 \\ -11 \\ 15 \end{pmatrix}$

**2** $\overrightarrow{AB} = \mathbf{b} - \mathbf{a} = \begin{pmatrix} 5 \\ 2 \end{pmatrix} - \begin{pmatrix} 4 \\ -2 \end{pmatrix} = \begin{pmatrix} 1 \\ 4 \end{pmatrix}$

**3** Magnitude $= \sqrt{3^2 + (-4)^2 + 12^2} = \sqrt{9 + 16 + 144} = \sqrt{169} = 13$

**4** Magnitude $= \sqrt{2^2 + (-3)^2 + 1^2} = \sqrt{4 + 9 + 1} = \sqrt{14}$

## Vectors (2)

**Q1** What is a *unit vector*?

**Q2** $\frac{1}{2}\mathbf{i} + \frac{1}{3}\mathbf{j} + m\mathbf{k}$ is a unit vector. Find the values of $m$.

**Q3** $\overrightarrow{AB} + \ldots + \overrightarrow{FG} = \overrightarrow{AG}$. What is the missing directed line segment?

**Q4** Calculate the distance $d$ between the points $(3, -1, 0)$ and $(-2, 5, 1)$.

**ANSWERS** ⟩⟩

**1** A vector with length (magnitude) of 1 unit.

**2** $\left(\dfrac{1}{2}\right)^2 + \left(\dfrac{1}{3}\right)^2 + m^2 = 1$

$\Rightarrow \quad \dfrac{1}{4} + \dfrac{1}{9} + m^2 = 1$

$\Rightarrow \qquad\qquad m^2 = \dfrac{23}{36}$

$\Rightarrow \qquad\qquad m = \pm\sqrt{\dfrac{23}{36}}$

**3** $\overrightarrow{BF}$

**4** $d = \sqrt{(x_2 - x_1)^2 + (y_2 - y_1)^2 + (z_2 - z_1)^2}$

$\quad = \sqrt{(-2 - 3)^2 + (5 + 1)^2 + (1 - 0)^2}$

$\quad = \sqrt{25 + 36 + 1} = \sqrt{62}$

# Vectors (3)

**Q1** What can you say about the vectors $\mathbf{a} = \begin{pmatrix} 2 \\ 3 \\ 4 \end{pmatrix}$ and $\mathbf{b} = \begin{pmatrix} 4 \\ 6 \\ 8 \end{pmatrix}$?

**Q2** A is the point (2, −3, 4) and B is (5, −9, 13). Find the coordinates of the point P that divides AB in the ratio 2 : 1.

**Q3** Show that the points P (3, 2, 6), Q (5, −2, 10) and R (9, −10, 18) are collinear.

**Q4** For the points in Q3, find the ratio in which Q divides PR.

**ANSWERS** ▶▶

**1** Vectors **a** and **b** are either parallel or collinear as **b** = 2**a**.

**2** Use the section formula.  A (2, −3, 4)   B (5, −9, 13)

2 : 1

P is $\left(\dfrac{1 \times 2 + 2 \times 5}{3}, \dfrac{1 \times (-3) + 2 \times (-9)}{3}, \dfrac{1 \times 4 + 2 \times 13}{3}\right)$

= (4, −7, 10).

**3** $\overrightarrow{PQ}$ = **q** − **p** = $\begin{pmatrix} 5 \\ -2 \\ 10 \end{pmatrix} - \begin{pmatrix} 3 \\ 2 \\ 6 \end{pmatrix} = \begin{pmatrix} 2 \\ -4 \\ 4 \end{pmatrix}$

$\overrightarrow{QR}$ = **r** − **q** = $\begin{pmatrix} 9 \\ -10 \\ 18 \end{pmatrix} - \begin{pmatrix} 5 \\ -2 \\ 10 \end{pmatrix} = \begin{pmatrix} 4 \\ -8 \\ 8 \end{pmatrix}$

As $\overrightarrow{QR}$ = 2$\overrightarrow{PQ}$, $\overrightarrow{QR}$ and $\overrightarrow{PQ}$ have the same direction.
As Q is a common point, then P, Q and R are collinear.

**4** By inspecting the above results, Q divides PR in the ratio 1 : 2.

## Vectors (4)

**Q1** Find the value of **a.b** when |**a**| = 6, |**b**| = 5 and $\theta$ = 30°, where $\theta$ is the angle between **a** and **b**.

**Q2** Find the value of **a.b** when $\mathbf{a} = \begin{pmatrix} 4 \\ 0 \\ -2 \end{pmatrix}$ and $\mathbf{b} = \begin{pmatrix} -3 \\ 5 \\ -7 \end{pmatrix}$.

**Q3** Find the value of **v.w** when **v** = 3**i** − 2**j** + **k** and **w** = **i** + 4**j** − 2**k**.

**Q4** Find the size of the angle between **v** and **w** in Q3.

**ANSWERS** ⟩⟩

**1**  $\mathbf{a.b} = |\mathbf{a}| \times |\mathbf{b}| \times \cos\theta$

$= 6 \times 5 \times \cos 30°$

$= 30 \times \dfrac{\sqrt{3}}{2}$

$= 15\sqrt{3}$

**2**  $\mathbf{a.b} = a_1 b_1 + a_2 b_2 + a_3 b_3$

$= 4 \times (-3) + 0 \times 5 + (-2) \times (-7)$

$= -12 + 0 + 14 = 2$

**3**  $\mathbf{v} = \begin{pmatrix} 3 \\ -2 \\ 1 \end{pmatrix}, \mathbf{w} = \begin{pmatrix} 1 \\ 4 \\ -2 \end{pmatrix}$

and hence $\mathbf{v.w} = 3 \times 1 + (-2) \times 4 + 1 \times (-2)$

$= 3 - 8 - 2 = -7$

**4**  $|\mathbf{v}| = \sqrt{3^2 + (-2)^2 + 1^2} = \sqrt{14}$

$|\mathbf{w}| = \sqrt{1^2 + 4^2 + (-2)^2} = \sqrt{21}$

$\cos\theta = \dfrac{\mathbf{v.w}}{|\mathbf{v}||\mathbf{w}|} = \dfrac{-7}{\sqrt{14} \times \sqrt{21}} = -0.408$

$\Rightarrow \theta = 114°$

## Vectors (5)

**Q1** What is the value of **a.b** when **a** and **b** are perpendicular?

**Q2** Evaluate **a.(a + b)** where |**a**| = 5, |**b**| = 6 and the angle between **a** and **b** is 60°.

**Q3** P and Q have position vectors $\begin{pmatrix} 2 \\ -1 \\ 3 \end{pmatrix}$ and $\begin{pmatrix} 4 \\ -3 \\ 1 \end{pmatrix}$, respectively.

Find the length of PQ.

**Q4** For what value of $m$ are the vectors $\begin{pmatrix} 2 \\ 1 \\ 4 \end{pmatrix}$ and $\begin{pmatrix} 2 \\ -5 \\ m \end{pmatrix}$ perpendicular?

**ANSWERS** ⟩⟩

**1** If **a** and **b** are perpendicular, then **a.b** = 0.

**2** a.(a + b) = a.a + a.b

$$= 5 \times 5 \times \cos 0° + 5 \times 6 \times \cos 60°$$

$$= 25 \times 1 + 30 \times 0.5 = 40$$

**3** $\overrightarrow{PQ} = \mathbf{q} - \mathbf{p} = \begin{pmatrix} 4 \\ -3 \\ 1 \end{pmatrix} - \begin{pmatrix} 2 \\ -1 \\ 3 \end{pmatrix} = \begin{pmatrix} 2 \\ -2 \\ -2 \end{pmatrix}$

$$\Rightarrow PQ = \sqrt{2^2 + (-2)^2 + (-2)^2} = \sqrt{12}$$

**4** $\begin{pmatrix} 2 \\ 1 \\ 4 \end{pmatrix} . \begin{pmatrix} 2 \\ -5 \\ m \end{pmatrix} = 2 \times 2 + 1 \times (-5) + 4 \times m$

$$= 4 - 5 + 4m = 4m - 1 = 0 \text{ when vectors are perpendicular.}$$

Hence $m = \dfrac{1}{4}$.

**_Exam_ tip:** Make sure you can use the formula $\cos\theta = \dfrac{\mathbf{a.b}}{|\mathbf{a}||\mathbf{b}|}$ to find the size of the angle between two vectors.

# Further calculus (1)

**Q1** Differentiate $3\cos x - 2\sin x$.

**Q2** Differentiate $(3x + 1)^7$.

**Q3** Differentiate $2\cos 5x$.

**Q4** Differentiate $\cos^3 x$.

**ANSWERS** ▶▶

**1** $-3\sin x - 2\cos x$

**2** $7(3x + 1)^6 \times 3 = 21(3x + 1)^6$ (Use the chain rule.)

**3** $2 \times (-5\sin 5x) = -10\sin 5x$ (Use the formulae list.)

**4** Think of $\cos^3 x$ as $(\cos x)^3$.

Hence the derivative is $3(\cos x)^2 \times (-\sin x) = -3\cos^2 x \sin x$.

# Further calculus (2)

**Q1** Find $\int(3x + 7)^9\,dx$.

**Q2** Find $\int\sin(3x - 2)\,dx$.

**Q3** Evaluate $\int_0^{\pi/2} (\sin 2x + \cos 2x)\,dx$.

**Q4** Calculate the area beneath the curve
$y = \cos 3x$ from $x = 0$ to $x = \dfrac{\pi}{6}$.

**ANSWERS** ❯❯

**1** $\int (3x + 7)^9 dx = \dfrac{1}{3 \times 10}(3x + 7)^{10} + C$

$$= \dfrac{1}{30}(3x + 7)^{10} + C$$

**2** $\int \sin(3x - 2)dx = -\dfrac{1}{3}\cos(3x - 2) + C$ (Use the formulae list.)

**3** $\displaystyle\int_0^{\pi/2} (\sin 2x + \cos 2x)\, dx = \left[ -\dfrac{1}{2}\cos 2x + \dfrac{1}{2}\sin 2x \right]_0^{\pi/2}$

$$= \dfrac{1}{2}\Big[ (-\cos \pi + \sin \pi) - (-\cos 0 + \sin 0) \Big]$$

$$= \dfrac{1}{2}(1 + 0 + 1 - 0) = 1$$

**4** Area $= \displaystyle\int_0^{\pi/6} \cos 3x\, dx = \left[ \dfrac{1}{3}\sin 3x \right]_0^{\pi/6}$

$$= \dfrac{1}{3}\left( \sin \dfrac{\pi}{2} - \sin 0 \right)$$

$$= \dfrac{1}{3}(1 - 0) = \dfrac{1}{3}$$

# Further calculus (3)

**Q1** Given that $f(x) = (2x - 1)^3$, find the value of $f'(2)$.

**Q2** If $\int\limits_0^b \cos x \, dx = 1$, find the value of $b$ if $0 \le b \le 2\pi$.

**Q3** Find $\int \cos^2 x \, dx$.

**Hint:** refer to Q1 of the section Compound and multiple angles (4).

**Q4** Evaluate $\int\limits_1^4 \sqrt{t} \, dt$.

**ANSWERS** ⟩⟩

**1** $f'(x) = 3(2x - 1)^2 \times 2$

$\phantom{f'(x)} = 6(2x - 1)^2$

$\Rightarrow f'(2) = 6 \times (2 \times 2 - 1)^2$

$\phantom{\Rightarrow f'(2)} = 6 \times 3^2 = 54$

**2** $\displaystyle\int_0^b \cos x \, dx = [\sin x]_0^b = \sin b - \sin 0 = \sin b = 1 \Rightarrow b = \dfrac{\pi}{2}$

**3** $\displaystyle\int \cos^2 x \, dx = \int \left( \dfrac{1 + \cos 2x}{2} \right) dx$

$\phantom{\int \cos^2 x \, dx} = \dfrac{1}{2} \int (1 + \cos 2x) dx$

$\phantom{\int \cos^2 x \, dx} = \dfrac{1}{2} \left( x + \dfrac{1}{2} \sin 2x \right) + C$

**4** $\displaystyle\int_1^4 \sqrt{t} \, dt = \int_1^4 t^{\frac{1}{2}} dt = \left[ \dfrac{2}{3} t^{\frac{3}{2}} \right]_1^4$

$\phantom{\int_1^4 \sqrt{t} \, dt} = \left[ \dfrac{2\sqrt{t^3}}{3} \right]_1^4$

$\phantom{\int_1^4 \sqrt{t} \, dt} = \left( \dfrac{2\sqrt{4^3}}{3} - \dfrac{2\sqrt{1^3}}{3} \right)$

$\phantom{\int_1^4 \sqrt{t} \, dt} = \dfrac{16}{3} - \dfrac{2}{3} = \dfrac{14}{3}$

# Further calculus (4)

**Q1** Evaluate $\int_{-2}^{2} (x + 1)^2 \, dx$.

**Q2** Find $\int \left( \dfrac{dx}{\sqrt{x + 5}} \right)$.

**Q3** Find $\int (x + 5\cos 3x) \, dx$.

**Q4** Given that $f(x) = 5x^2 - 3\sin x$, find $f'(0)$.

**ANSWERS** ▸▸

**1** $\int\limits_{-2}^{2} (x + 1)^2 \, dx = \left[\frac{1}{3}(x + 1)^3\right]_{-2}^{2}$

$$= \frac{1}{3}(2 + 1)^3 - \frac{1}{3}(-2 + 1)^3$$

$$= \frac{1}{3}[27 - (-1)] = \frac{28}{3}$$

**2** $\int\left(\frac{dx}{\sqrt{x + 5}}\right) = \int(x + 5)^{-\frac{1}{2}} \, dx$

$$= 2(x + 5)^{\frac{1}{2}} + C$$

**3** $\int(x + 5\cos 3x) \, dx = \frac{1}{2}x^2 + 5 \times \frac{1}{3}\sin 3x + C$

$$= \frac{1}{2}x^2 + \frac{5}{3}\sin 3x + C$$

**4** $f(x) = 5x^2 - 3\sin x$

$\Rightarrow f'(x) = 10x - 3\cos x$

$\Rightarrow f'(0) = 0 - 3\cos 0 = -3$

***Exam tip:*** When you are differentiating or integrating sine or cosine functions, always look up the formulae list.

## Logarithms (1)

**Q1** Express $m^n = p$ in logarithmic form.

**Q2** Express $\log_x y = z$ in index form.

**Q3** Write down the values of $\log_a a$ and $\log_a 1$.

**Q4** Evaluate $\log_2 \frac{1}{16}$.

**ANSWERS** ⟩⟩

**1** $m^n = p \Rightarrow \log_m p = n$

**2** $\log_x y = z \Rightarrow y = x^z$

**3** $\log_a a = 1, \log_a 1 = 0$

**4** Let $\log_2 \dfrac{1}{16} = x$

$\Rightarrow \qquad 2^x = \dfrac{1}{16} = \dfrac{1}{2^4} = 2^{-4}$

$\Rightarrow \qquad x = -4$

# Logarithms (2)

**Q1** Express $\log 25 + \log 4 - \log 20$ as the logarithm of a single number.

**Q2** If $\log a + 3 \log 2 = \log 56$, find $a$.

**Q3** Solve the equation $3^x = 8$, giving your answer correct to one decimal place.

**Q4** If the graph of $y = \log_a x$ passes through the point $(9, 2)$, find $a$.

**ANSWERS** ▶▶

**1**  $\log 25 + \log 4 - \log 20 = \log\left(\dfrac{25 \times 4}{20}\right)$

$$= \log\left(\dfrac{100}{20}\right) = \log 5$$

**2**  $\log a + 3\log 2 = \log a + \log 2^3$

$$= \log(a \times 2^3)$$

$$= \log 8a$$

$\Rightarrow \qquad 8a = 56 \Rightarrow a = 7$

**3**  $3^x = 8$

$\Rightarrow \ \log 3^x = \log 8$

$\Rightarrow x\log 3 = \log 8$

$\Rightarrow \qquad x = \dfrac{\log 8}{\log 3}$

$$= 1 \cdot 9 \quad \text{(to one decimal place)}$$

**4**  Substitute $(9, 2)$ into $y = \log_a x$

$\Rightarrow \ 2 = \log_a 9$

$\Rightarrow \ a^2 = 9$

$\Rightarrow \ a = 3$

# Logarithms (3)

**Q1** Evaluate $\log_{\sqrt{2}} 8$.

**Q2** Evaluate $\dfrac{\log_b 27}{\log_b 3}$.

**Q3** Given that $\log_{10} x = 3\log_{10} y + \log_{10} 2$, express $x$ in terms of $y$.

**Q4** The number of bacteria $N(t)$ in a sample after $t$ hours is given by the formula $N(t) = 50e^{1.5t}$.

How many bacteria were there at the start?

**ANSWERS ▸▸**

1 Let $\log_{\sqrt{2}} 8 = x$

$\Rightarrow \quad (\sqrt{2})^x = 8$

$\Rightarrow \quad (2^{\frac{1}{2}})^x = 2^3$

$\Rightarrow \quad \frac{1}{2}x = 3$

$\Rightarrow \quad x = 6$

2 $\dfrac{\log_b 27}{\log_b 3} = \dfrac{\log_b 3^3}{\log_b 3} = \dfrac{3\log_b 3}{\log_b 3} = 3$

3 $\log_{10} x = 3\log_{10} y + \log_{10} 2$

$\qquad = \log_{10} y^3 + \log_{10} 2$

$\qquad = \log_{10} 2y^3$

$\Rightarrow \quad x = 2y^3$

4 At the start, $t = 0$, and hence $N(0) = 50e^{1.5 \times 0}$

$\qquad\qquad\qquad\qquad\quad = 50e^0$

$\qquad\qquad\qquad\qquad\quad = 50 \times 1$

$\qquad\qquad\qquad\qquad\quad = 50.$

# Logarithms (4)

**Q1** Solve the equation $e^{3x} = 80$ for $x$.

**Q2** Express $Y = 0 \cdot 5X + 3$ in the form $y = ax^n$,
where $Y = \log_{10} y$ and $X = \log_{10} x$.

**Q3** If the graph of $y = \log_5 (x - a)$ passes through the point $(6, 1)$, find $a$.

**Q4** Express $3 + \log_2 5$ as the logarithm of a single number.

**ANSWERS** »

**1** $e^{3x} = 80$

$\Rightarrow \quad \ln e^{3x} = \ln 80$

$\Rightarrow \quad 3x \ln e = \ln 80$

$\Rightarrow \qquad x = \dfrac{\ln 80}{3} = 1{\cdot}46$

**2** $Y = 0{\cdot}5X + 3$

$\Rightarrow \qquad \log_{10} y = 0{\cdot}5 \log_{10} x + 3$

$\qquad\qquad\quad = \log_{10} x^{0{\cdot}5} + \log_{10} 1000$

Hence $\log_{10} y = \log_{10} 1000 x^{0{\cdot}5}$

$\Rightarrow \qquad\qquad y = 1000 x^{0{\cdot}5}$

**3** $y = \log_5 (x - a)$

$\Rightarrow \quad 5^y = x - a$

$\Rightarrow \quad 5^1 = 6 - a$

$\Rightarrow \quad a = 6 - 5 = 1$

**4** $3 + \log_2 5 = 3 \log_2 2 + \log_2 5$

$\qquad\qquad\quad = \log_2 2^3 + \log_2 5 = \log_2 (2^3 \times 5) = \log_2 40$

***Exam* tip 1**: It will help you to solve many problems in logarithms if you are able to convert from logarithmic form to index form and vice versa, that is $\log_a b = c \Leftrightarrow a^c = b$.

***Exam* tip 2**: As a guide, remember that $10^3 = 1000 \Leftrightarrow \log_{10} 1000 = 3$.

## The auxiliary angle (1)

**Q1** If $k\sin a° = 3$ and $k\cos a° = 4$, find $k$.

**Q2** For the equations in Q1, find $a$, $\quad 0 < a < 360$.

**Q3** If $2\sin x° + 3\cos x° = k\sin(x - a)°$, find the values of $k\sin a°$ and $k\cos a°$.

**Q4** For the equation in Q3, find $k$ and $a$.

**ANSWERS** ▶▶

**1** $k = \sqrt{3^2 + 4^2} = 5$

**2** $\tan a° = \dfrac{3}{4}$ and $a$ lies in the first quadrant,

and hence $a = \tan^{-1} 0\cdot 75 = 36\cdot 9$.

*Note: $a$ is in the first quadrant because sin and cos are both positive.*

**3** $2\sin x° + 3\cos x° = k\sin(x - a)°$

$$= k\sin x° \cos a° - k\cos x° \sin a°$$

Hence $\quad k\sin a° = -3$ and $k\cos a° = 2$.

**4** $k = \sqrt{(-3)^2 + 2^2} = \sqrt{13}$

$\tan a° = -\dfrac{3}{2} \Rightarrow a = 303\cdot 7$

*Note: $a$ is in the fourth quadrant because sin is negative and cos is positive.*

# The auxiliary angle (2)

**Q1** For $\cos x - \sin x = k\cos(x - a)$, $0 \le a \le 2\pi$, find the values of $k\sin a$ and $k\cos a$.

**Q2** For the equation in Q1, find $k$ and $a$.

**Q3** Using the information in Q1 and Q2, find the maximum value of $\cos x - \sin x$.

**Q4** What is the value of $x$ for which the maximum value in Q3 occurs?

**ANSWERS** ▶▶

**1** $\cos x - \sin x = k\cos(x - a)$

$$= k\cos x \cos a + k\sin x \sin a$$

Hence $k\sin a = -1$ and $k\cos a = 1$.

**2** $k = \sqrt{(-1)^2 + 1^2} = \sqrt{2}$

$\tan a = -\dfrac{1}{1} = -1 \implies a = \dfrac{7\pi}{4}$ radians

**3** The maximum value of $\cos x - \sin x$ is $\sqrt{2}$.

**4** This occurs when $\cos\left(x - \dfrac{7\pi}{4}\right) = 1$

$\implies \qquad\qquad\qquad x - \dfrac{7\pi}{4} = 0$

$\implies \qquad\qquad\qquad\qquad x = \dfrac{7\pi}{4}.$

# The auxiliary angle (3)

**Q1** What is the minimum value of $5\cos(x + 28)° + 2$?

**Q2** What is the value of $x$ for which the minimum value in Q1 occurs?

**Q3** If $\sin x - \sqrt{3}\cos x = 2\sin\left(x - \dfrac{\pi}{3}\right)$,

solve the equation $\sin x - \sqrt{3}\cos x = \sqrt{2}$, $\quad 0 \le x \le 2\pi$.

**Q4** For what values of $x$ does $4\cos(x - 25)° = 0$, $\quad 0 \le x \le 360$?

**ANSWERS** ▶▶

**1** The minimum value is $-5 + 2 = -3$.

**2** It occurs when $\cos(x + 28)° = -1$, that is when

$x + 28 = 180$

$\Rightarrow \quad x = 152$.

**3** $\sin x - \sqrt{3} \cos x = \sqrt{2}$

$\Rightarrow 2\sin\left(x - \dfrac{\pi}{3}\right) = \sqrt{2}$

$\Rightarrow \quad \sin\left(x - \dfrac{\pi}{3}\right) = \dfrac{\sqrt{2}}{2} = \dfrac{1}{\sqrt{2}}$

Hence $\quad x - \dfrac{\pi}{3} = \dfrac{\pi}{4}$ or $\dfrac{3\pi}{4}$

$\Rightarrow \qquad\qquad x = \dfrac{7\pi}{12}$ or $\dfrac{13\pi}{12}$

**4** $4\cos(x - 25)° = 0$ when $x - 25 = 90$ or $270$,

that is when $x = 115$ or $295$.

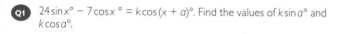

## The auxiliary angle (4)

**Q1** $24 \sin x° - 7 \cos x° = k \cos(x + a)°$. Find the values of $k \sin a°$ and $k \cos a°$.

**Q2** For the equation in Q1, find $k$ and $a$.

**Q3** From Q2 solve the equation $24 \sin x° - 7 \cos x° = 15$, $0 \leq x \leq 360$.

**Q4** What is the minimum value of $\dfrac{1}{24 \sin x° - 7 \cos x°}$?

**ANSWERS** ▶▶

**1** $24 \sin x° - 7 \cos x° = k \cos(x + a)°$

$$= k \cos x° \cos a° - k \sin x° \sin a°$$

Hence $k \sin a° = -24$ and $k \cos a° = -7$.

**2** $k = \sqrt{(-24)^2 + (-7)^2} = \sqrt{625} = 25$

$\tan a° = \dfrac{-24}{-7} = \dfrac{24}{7}$

$\Rightarrow \quad a = 253·7$

*Note:* $a$ is in the third quadrant because sin is negative and cos is negative.

**3** $24 \sin x° - 7 \cos x° = 15$

$\Rightarrow \quad 25 \cos(x + 253·7)° = 15$

$\Rightarrow \quad \cos(x + 253·7)° = 0·6$

Hence $\quad x + 253·7 = 53·1$ or $306·9$

$\Rightarrow \qquad\qquad x = 53·2$ or $159·4$

**4** $\dfrac{1}{25}$

*Exam tip:* To express $a \cos x + b \sin x$ in the form $k \cos(x \pm a)$ and $k \sin(x \pm a)$, start by using the appropriate formula from $\cos(x \pm a)$ and $\sin(x \pm a)$, and then find values for $k \sin a$ and $k \cos a$.